THE HOLLY BLUE B

CW00664114

By Ken Willmott F

Illustrations by Dr Tim

Photographs by the author

This booklet is sponsored by the following branches of Butterfly Conservation:-

Butterfly Conservation
**HERTFORDSHIRE &
MIDDLESEX
BRANCH**

HAMPSHIRE AND ISLE OF WIGHT BRANCH

INDEX

INTRODUCTION

The Holly Blue (*Celastrina argiolus Linn.* 1758)

It is more than appropriate that a booklet on the Holly Blue be produced in 1999, the year after 'Gardening for butterflies' campaign began. The Holly Blue is very much a garden butterfly. It is not just an occasional visitor at nectar sources, but a frequently breeding species. It particularly breeds on the holly and the ivy, both of which are found in many garden situations.

The Holly Blue is a most interesting and delightful species, a harbinger of spring, one of the first butterflies to emerge from its overwintering pupa. It often coincides and flies with the Speckled Wood (*Pararge aegeria*) and Small White (*Pieris rapae*) in April and is perhaps our only butterfly, together with the Brimstone (*Gonepteryx rhamni*) hibernating amongst Ivy, whose appearance on a Christmas card would not be out of place!

This booklet is essentially an account of a long term study of the Holly Blue, undertaken within the confines and immediate vicinity of John Innes Recreation Ground, Merton Park, a beautiful ornamental garden in a South London suburb. Here, to this day, a strong population flourishes which can be identified as a rarely found 'core' population. This term describes a population that remains relatively stable each year in comparison with those more nomadic populations found in the general countryside, which can fluctuate dramatically from year to year as the transect figures in the appendices clearly show. The transect figures from different regions show variations, e.g. a comparison of Durlston Country Park (Dorset coast) with Ashford Hangers and Wendleholme (Hampshire) for 1996-97 illustrates that population fluctuation is not always synchronous. An extreme example comes from Surrey where 1997 transect figures showed a decrease of 80% from those of 1996, quite the reverse of the Dorset coast figures.

To understand the principles and foundation of a 'core' population a brief account of the history and possible origins of the colony is worthwhile.

The South London gardens at Merton Park were opened in 1909 and named after its founder, John Innes. He was a city businessman, property developer and much respected Victorian landowner. In 1867 when John Innes purchased the Merton Estate it was a rambling north Surrey parish with a small population of approximately two thousand. In the late 1890's London expansion had reached Merton and John Innes new suburb was based on an old and primarily agricultural Surrey village. He was the creator of the new Merton Park (although he preferred the name Merton Manor) where the 'core' population of the Holly Blue still flies in the original 'garden suburb' as it was then called.

2

Three long straight roads were built in the 1870's and thousands of holly trees and hedges were planted. The holly became the trade mark of Merton Park, the favourite tree of John Innes and a symbol of the Innes 'clan'. A significant number of its main larval foodplant were thus condensed into a small area, and these were supplemented by ivy, as a natural course, scrambling over many old and newly constructed walls. As the garden suburb became established from developed farmland, conditions became increasingly ideal for a steady increase in the Holly Blue population as its larval foodplants matured.

John Innes died in August 1904 before the establishment, in 1910, of his Horticultural Research Institution. This finally moved to Norwich, East Anglia in 1967. Thirty years after his death his renowned potting composts gave him posthumous fame. The institute, as well as the ornamental gardens, were founded in his memory. The local twelfth century church which he frequented during his life and the churchyard where he is buried also supports a quantity of holly and ivy and the Holly Blue can still be seen here. **The importance of churchyards for the Holly Blue cannot be stressed enough**. For example it has been recorded in 78% of Oxfordshire churchyards. The Merton church could have been the original source for the Holly Blue in the 'garden suburb'.

As a further memorial to John Innes, a society was formed in his name and parts of his original garden suburb declared a 'Conservation Area' designated to serve as a focal point from which environmental care and improvement should flow outwards. The John Innes Society finances tree planting in order to retain and extend the green character of the district. Strangely, the name Merton is derived from 'town on the marsh', a somewhat less than attractive habitat for the Holly Blue.

Even in Merton Park today the strong status of the Holly Blue seems assured, despite several setbacks having perhaps temporarily lowered its numbers. This is discussed further in the conservation section. Dr Jeremy Thomas describes his amazement on entering the gardens in the 1980's in the opening account of the Holly Blue in his classic work - *The Butterflies of Britain and Ireland* (Dorling Kindersley 1991) which is so admirably illustrated (including the important host-specific parasite) by Richard Lewington.

Sections which follow are mostly from the pages of my field notebook, the result of many hours observations. Although some may seem dated, they are still relevant to current Holly Blue populations. The transect figures in tables 2-5 are the results of regular walks by volunteers, counting butterflies seen five metres ahead of them and are a highly useful means of monitoring population fluctuations. Some recorders imagine themselves in a moving five metre box and record all butterflies that they see within the box. They walk a regular route at a steady pace throughout the season, when there are suitable amounts of sunshine and temperatures amenable to butterfly flight.

The late James Petiver, who produced *Papilionum Britanniae* (1717), the first publication devoted wholly to British butterflies, does not seem to list this species since John Ray first described it in 1710. There is some confusion, perhaps because of the marked sexual dimorphism shown by the Holly Blue, as with the Brown Hairstreak (*Thecla betulae*), whose distinctly marked sexes were initially described as separate species. The male Holly Blue was called the Blue Speckt butterfly whilst the female, the Blue Speckt butterfly with Black Tipps. Over fifty years passed without further reference until John Berkenhout listed it in 1769, but only under its Linnaean scientific name. He was the first to adopt the Linnaean methods. In 1775 Moses Harris included it in *The Aurelians Pocket Companion* as the Azure Blue. Twenty years later it was given another name, Wood Blue by William Lewin, presumably because it was the only blue butterfly seen regularly in a woodland habitat. In these days of wide grassy rides and timber extraction roads habitat is also created for the Common Blue (*Polyommatus icarus*). The Azure Blue was adopted by many authors and admirably depicts the brilliant sky blue colour that adorns the upper surface of the wings, especially the male.

The Reverend Francis Orpen Morris in his 1853 work *A History of British Butterflies* appears to be amongst few, and perhaps the first, to use the current English name of Holly Blue, before Kane in 1885 and Richard South in 1906. It has been used ever since. The Reverend Morris recognised the Holly Blue as being double brooded, but only mentions Holly and Buckthorn (*Rhamnus* sp.) as larval foodplants with no reference to its most important association with Ivy (*Hedera helix*).

Recent history refers to the remarkable nature of its population fluctuations. In some recent years the Holly Blue became so common as to warrant space in national and local press. Such noticeably high populations are not a new event and thus may not be due to global warming. Frederick William Frohawk (1934) remarks on the highs and lows of Holly Blue populations. He relates how from 1900 it became increasingly common, especially in suburban gardens, until 1931 when it became relatively scarce, only to increase dramatically again three years later in areas where it was once uncommon.

In FW Frohawk's 1938 book on varieties he has a fine illustration of a female Holly Blue completely lacking the black spots on the underside and calls it *ab. parvipunctata*. He captured this second brood specimen on 27 July 1929 in Sutton, Surrey, close to the London Borough of Merton where this current study was undertaken. This is where FWF settled in his last home from 1929 until his death in 1946, after having to sell reluctantly his magnificent butterfly collection in order to help purchase the Sutton property.

4

Variation in the Holly Blue is scarce and more or less confined to the black spotting on the underside of the wings. There are obsolete varieties, with some or all of the spots missing, elongated spots, such as ab. subtus-radiata, or sometimes a variation in the shade of the blue colour. Where this is extreme it could be a pathological variety, where the wing scales are deformed. Perhaps the best known pathological varieties being those of the Meadow Brown (*Maniola jurtina*) where there are distinct white patches on their wings, rarely covering the entire wing surfaces.

The Holly Blue exhibits both **sexual** and **seasonal** dimorphism. Females have extensive areas of black on the apex and leading edge of the forewings and to a lesser extent the hindwings. This is reduced to a minimal apical extension and thin line along the outer margin in the male. Where it is double brooded the summer females are more heavily suffused in black than their spring predecessors.

CURRENT DISTRIBUTION

In Great Britain it is very much a species of southern distribution, becoming less frequent north of the Midlands. It is found throughout Wales and Ireland, including the Isle of Man. In Scotland it is said to be a rare vagrant of southernmost areas. This situation could change as a result of predicted climatic change. Somewhat isolated populations in the north-west have a single brood in spring, with the exception of during very warm summers, when second brood examples have been recorded, as they have, for example, in Yorkshire. The forthcoming Butterfly Conservation *Millenium Atlas* will give an up to date distribution map for the Holly Blue in the British Isles. It is a wide ranging species throughout Europe, North Africa, Turkey, Central Asia, Japan and North America (Tolman 1997), where it is known as the Spring Azure. One of the reasons it is a relatively common and widespread species throughout the world is that over **two hundred** larval foodplants have been recorded, with over **fifty** from the United States alone.

HABITS AND LIFE - HISTORY

The under surface of the wings of the Holly Blue are a very pale sky blue, with a series of black spots on both fore and hind wings. Few people realise the remarkable qualities of camouflage that the undersurface of the wings affords this species. It takes rest amongst evergreen leaves, overnight, during periods of dull weather and at its most vulnerable time, having recently emerged from the pupa and still drying its wings before flight. These are often leaves of the holly and the ivy, whose waxy leaf surfaces are smooth and shining and reflect daylight from every angle as one's eyes scour them for possible resting adults. The pale underside blends perfectly with the mass of bright, shining leaves.

Despite the area of study being mostly confined to suburban parkland, the Holly Blue is a butterfly of very varied habitat, particularly in good years when it expands from its 'core' populations. Individuals have been found on heathland, amongst the Silver-studded Blues (*Plebejus argus*), in coppiced woodland, amongst the Heath Fritillaries (*Mellicta athalia*), and along the coastal fringe amongst the Lulworth Skippers (*Thymelicus acteon*). It is basically a woodland butterfly, as indeed are its principal larval foodplants and the South London parkland closely 'mimics' woodland with its towering hollies and other mature evergreen shrubs. Even Peacocks (*Inachis io*) and Commas (*Polygonia c-album*) establish territories each spring along the tarmac paths that criss-cross the well manicured gardens. The Speckled Wood (*Pararge aegeria*) also breeds, along with the odd Orange Tip (*Anthocharis cardamines*) and vagrant passing Brimstone (*Gonepteryx rhamni*).

The majority of observations in this booklet have been made amongst a Holly Blue population utilising solely holly and ivy as its larval foodplant, but there have been observations in other localities of females depositing eggs on a wide range of interesting plants and shrubs. Ling (*Calluna* sp.) in heather country, Box (*Buxus* sp.) in a Surrey churchyard, Alder Buckthorn (*Rhamnus* sp.) in woodland and perhaps most surprisingly, Birds-foot Trefoil (*Lotus* sp.) in clay grassland, where Common Blue (*Polyommatus icarus*) females were depositing eggs alongside!

A list of recorded plants upon which female Holly Blues have been observed depositing eggs can be found in the Appendix. A word of caution here. Not all records confirm the actual location of a deposited egg by the observer and a female was once observed crawling and probing amongst the florets of *Buddleia davidii*. It appeared certain that an egg had been deposited after so much probing and positioning by the female. Despite this, no egg was located after an exhaustive breakdown of the suspected flower.

TIME OF APPEARANCE

The Holly Blue is double brooded in the south of its range and single brooded in the north, each brood flies for a period of approximately six weeks. In poor summers, the second brood in the south does not always comprise a full emergence of adults. 1970 was an example of a complete emergence of the second brood, with ivy flower buds covered in eggs and resulting larval numbers very high. In contrast, 1981 saw few individuals and banks of ivy were much less utilised. Such numerical fluctuation can clearly be seen from the transect figures in Tables 2-5.

In normal seasons the Holly Blue begins to emerge in mid-April. Even in the appalling spring of 1983 (the great Clouded Yellow (*Colias croceus*) year!) it was sighted on the 16 April (temperature 65°F/18°C) under the influence of a ridge of high pressure which resulted in two days respite from cool, unsettled weather. In 1997 they were on the wing during the last few days of March, which is exceptional, but may happen with increasing frequency if global warming is a reality.

The month of May, particularly the first week, is usually when the species is at the peak of its emergence. There are then fresh examples of both sexes to satisfy the needs of the photographer, together with pairing and egglaying activities in progress to satisfy the keen observer. Some of the latest dates I have seen the first brood is 12 June 1983 and 10 June 1975. The second brood, where it occurs, begins to emerge during the first half of July. In the summer of 1970 it was seen on 6 July and the great drought summer of 1976 on 5 July. It flies throughout the month of August and egglaying females can be seen during the first part of September, although they are then mostly rather weak and tatty in appearance.

In the lepidoptera literature, October specimens are often recorded as third brood examples, but I suspect they may be stragglers, late emergers of second brood individuals from eggs deposited late in the flight of the spring brood. By late July eggs are already found on the unopened flowers of the ivy, as well as first instar larvae. One of the latest dates I have seen the Holly Blue was 10 September 1983, a female in fair condition flying along a bank of ivy, amongst several final instar larvae. The lifespan of the Holly Blue is between 15-20 days (FW Frohawk), giving an earliest estimated day of emergence of the late example as 21 August 1983.

ADULT BEHAVIOUR

The Holly Blue is the only one of our blue butterflies which consistently flies above head height, in a similar manner to our Hairstreak species. Most high flying blue butterflies encountered in shrubby areas will be the Holly Blue. It does, on occasion, come down low enough to be photographed, often whilst basking in sunspots, nectaring (mainly with closed wings), or whilst females are busily engaged in egglaying activities, during which time they will often rest, especially if the sun has suddenly disappeared. On its reappearance they will often briefly open their wings before continuing with their activities.

Males are fond of alighting on damp patches on road surfaces or woodland tracks and also have a predilection for animal droppings. A male has been observed feeding from the droppings of a wood pigeon. Important sources of nectar for the first brood of the Holly Blue are the flowers of holly, Laurel, Laurestinus, Bay Tree, Forget-me-not, Spanish Hyacinth and away from its parkland habitat it has been seen feeding upon the flowers of Garlic Mustard. A favourite source for second brood individuals is undoubtedly the flowers of Snowberry, upon whose blooms several individuals can congregate. Buddleia davidii and Wisteria are also favourites and away from the parkland environment Bramble and Heather are particularly attractive. There must be many more flowers that attract the attention of the Holly Blue. The flowering of Lilac often coincides with the first brood of the Holly Blue, but despite its frequency in suburban gardens, it does not seem attractive.

The Holly Blue has a rapid flight, particularly the male, as they closely follow the contours of large holly trees and other evergreen shrubs or along sunny banks of clinging ivy. Although seemingly restless on sunny days they do settle in the early and latter parts of the day in particular. On settling, they immediately orientate themselves into the rays of the sun to gain maximum benefit, especially on chill April mornings. They bask with wings half open, never allowing them to lie flat. Both sexes bask in this manner, the more active males being observed more frequently. Most active females are engaged in egglaying activities particularly after periods of nectaring. Egglaying flight is slow and deliberate often taking some time to locate a suitable surface for the egg to be deposited.

Females can often be found on warmer mornings during the peak flight period, freshly emerged and inactive on an ivy leaf, especially along large south facing banks. Two individuals were located at 10.00 hrs., after a mild night (50°F/10°C) close together, one on the top side of an ivy leaf, the other deep inside the tangled roots. Both still had limp wings, probably having emerged from the pupa an hour or so earlier (28 April 1975). One female, discovered on the ivy early in the morning (14 May 1983) did not take flight until 14.20 hrs. despite suitable periods of sunshine, perhaps exemplifying the inactivity of the female. Perhaps she was awaiting discovery by a patrolling male?

The activity of the Holly Blue is greatly reduced when the wind is from a northerly or north-easterly direction. Spells of northerly winds can frequently occur during the flight period of the first brood, and can sometimes last for a period of 7-10 days, thus seriously curtailing the activities of the Holly Blue. Good years for the Holly Blue could possibly be influenced by a **lack** of northerly airstreams during the spring flight period, although there are many other factors which must determine a successful or unsuccessful year for this species. Even during lengthy sunny spells in cool temperatures they will often only fly in a few well sheltered locations, and usually only males take exercise, flying around to the sunny side of tall hollies. Here they spend lengthy periods basking in order to accumulate enough heat to raise their body temperatures enough to sustain flight. The clear, cold nights that often accompany northerly airstreams also inhibit the morning pupal emergence of adults from the ivy growths.

The male Holly Blue engages conspecific males, often during patrolling flights, in mildly aggressive aerial confrontations. However, they normally locate females by the sexual strategy known as 'patrolling', rather than as 'perching' and awaiting females to enter established territories. After having warmed themselves sufficiently on their overnight roosts they tirelessly patrol, investigating the banks of ivy and tall holly trees in the hope of locating a receptive female. Unreceptive females, that have already been mated, reject the advances of a persistent male by rapidly vibrating their wings and spinning away from the curving, probing male abdomen. Finally, if the male continues his amorous advances, she drops from her resting leaf and crawls away into the shade or under another leaf in order to confuse and discourage further attention.

The Holly Blue has been observed in copulation on several occasions, but the brief courtship flight leading up to successful pairing, less frequently. On 6 May 1981 a fine late afternoon followed a mostly cloudy day, but it was noticeably warmer than the previous few days. The cloud cleared at 16.30 hrs. A resting female was discovered by an actively searching male and she immediately led him from her resting place onto a suitably sited 'platform', in this case a large ivy leaf, part of an extensive growth covering a large expanse of brick wall. They paired at 17.13 hrs and were found still 'in cop' at 18.45 hrs. that evening. The pairing site was re-visited early next morning and the couple found parted, in the shade, on separate ivy leaves, in close proximity.

EGGLAYING

Eggs can be found, in situ, of both broods of the Holly Blue, firstly on well visited holly trees and secondly on banks of sunny ivy growth. The eggs on ivy are usually more numerous and easier to locate in contrast with those on holly, which are normally of more scattered distribution and often out of reach compared with the growth form of some condensed banks of ivy, which is often utilised by a large number of females. Even on holly, eggs are sometimes laid at low elevation and regularly trimmed hedge hollies, as long as flower production is evident, will also be used.

Females begin their egglaying activities from mid-morning and will continue very late in the day if the weather is suitable. They have been observed to commence egglaying at 16.00 hrs. after an improvement in the weather - a typical British spring or summer situation, the F.E.S. - Fine Evening Syndrome, so well known to butterfly enthusiasts! It is encouraging that the Holly Blue takes advantage!

They most frequently deposit their eggs at the base of unopened flower buds (both holly and ivy), but sometimes on the stems. They are very careful to position their eggs onto the tougher green sheath of the bud and not on the softer, expanding petal area which would eventually fall to the ground with the egg. The holly tree is dioecious (male and female flowers on separate trees) and the female does not distinguish between the sexes. The female will also select several holly varieties, such as *Ilex aquifolium* Handsworth New Silver (variegated) and the Hedgehog holly *ilex aquifolium ferox* but shows a preference for the traditional English holly.

From my earlier observations in the 1970's, it appeared that eggs deposited on male holly trees were unsuccessful, the resulting larva inevitably perishing, as a result of having no berries on which to feed. From half a dozen eggs marked out in situ and re-visited, the only two survivors were those on female holly trees, the larvae located and found burrowing their retractible heads into the developing drupe. It proved to be too small a sample (through lack of time to monitor a larger sample) and rather unfortunate that the four larvae on male trees, despite their initial success on the unopened flower buds, perished. In 1984 Dr Ernie Pollard, then with the Institute of Terrestrial Ecology, discovered larvae on male holly bushes at Woodhurst in Cambridgeshire, feeding on the soft terminal leaves. My earlier findings from the small sample had been unfortunately reported in the 1984 *Atlas of Butterflies in Britain and Ireland* (Thomas, Heath, Pollard). Thanks to Dr Pollard's work, this can now be corrected and the year following his observations larvae were readily found on male trees in the South London parkland.

The eggs of the second brood are often deposited on the ivy flowers when they are still rather small and unopened. Females have difficulty placing their eggs underneath the small compact florets. First brood females often have a contrasting problem. Sometimes a high proportion of the holly buds have already opened and flowered. Unopened buds on which to deposit eggs can be difficult to locate. This situation varies annually, depending on whether the holly flowering season is advanced or retarded. Holly Blue flight can be interrupted by a period of unsettled weather before the peak flowering of holly, when there are ample flower buds available. Such an event can have detrimental consequences for Holly Blue populations. In the poor spring of 1983 holly flowered prolifically 15-17 May. This was very noticeable through great bee activity on the flowers. Two days before, a

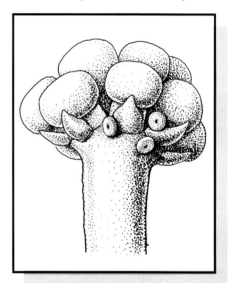

Figure 1: The eggs of the Holly Blue on the immature flower buds of ivy.

pairing and a female egglaying were recorded so there were no doubt many more eggs to be deposited despite the suddenly advanced holly flowers and the diminished number of closed flower buds available. This is another example of how a climatic situation could have possible implications for Holly Blue populations.

THE EGG

The egg is very pale blue in colour, although when first deposited it is a darker turquoise

The larva changes it's skin three times before transformation into the pupal stage (below right) and has three colour forms, with the green form being the most frequent. Note the shiny black head of the larva, only visible when it is active.

A female intent on depositing an egg on the unopened flower bud of the Ivy. Below shows an egg carefully positioned on the bud sheath of the Holly flower.

The urban habitat of the Holly Blue - John Innes Recreation Ground in full Spring bloom
and the churchyard with it's Holly and Ivy, foodplant for the larval stage.

A thermoregulating male warms himself on an ivy leaf in the morning sun, before actively seeking a mate. The spring brood of the female (below) is less heavily marked in black than summer brood females (front cover).

green. The colour changes quite quickly. The eggs hatch between 10-16 days (see Table 1) and the chorion (eggshell) remains uneaten. From the egg hatching figures it is interesting to note some possible temperature variables. In the spring of 1974 both March and April were anticyclonic months (cool nights) whereas March, April and May in 1983 were very much cyclonic with milder nights. All the data in the table is derived from wild eggs marked and re-visited, with the exception of 1977 data which was from a captive egg, kept out of doors and close to the original habitat.

TABLE 1 - **Egg Hatching Period**

YEAR	DATE EGG LAID	DATE EGG HATCHED	PERIOD
1974	10 May	20 May	10 Days
1977	7 May	22 May	15 Days
1979	14 May	26 May	12 Days
1983	15 May	31 May	16 Days

THE LARVA

The first instar larva of the Holly Blue (2-3 mm.) is attacked by a large ichneumonid wasp - *Listrodromus nycthemerus*. This host-specific parasite can be readily observed during the first two weeks of August, flying from one spray of ivy flower buds to another, in search of its prey. It is rarely seen in May or June probably because first brood larvae on holly are more widely scattered than they are on the banks of ivy. The adult parasite emerges from the pupa of the Holly Blue, but it is uncertain at what time the parasite emerges in a wild situation. It may be during May and August, thus overwintering inside the pupal shell or perhaps it emerges September/October and spends the winter as an adult. I know of no studies into this, but the former seems more logical to me.

On several occasions the final instar larva has been found with ants in attendance. On one occasion (17 September 1977) close to the Sussex coast on an old ivy-covered wooden barn several larvae were receiving attention

Figure 2: The host specific parasite Listrodromus nycthemerus seeking a first instar larva.

11

from red ants of the genus *Myrmica*. Larvae were not always situated low down on the bank of ivy, so ants had foraged some distance from their nests to locate the attractive secretions of the Holly Blue larvae. In the suburban parkland environment larvae have been found similarly attended, but by black ants of the genus *Lasius*. In this instance only larvae feeding low down on the south facing ivy were receiving attention, those at higher elevation were unattended.

From marking and re-locating, larvae have been found on female holly trees where they feed on the immature contents of green berries. The larvae wraps itself around the outside of the berry and burrows its specialised head and front segments inside, completely devouring the contents, leaving a tough empty green shell. These eventually wither and turn brown. A group of such withered berries can indicate the presence of a nearby larva. Larvae on male holly trees eventually feed on the new, tender terminal leaves and can be located by the presence of transparent 'windows' on leaves, made by the larvae only devouring the leaf cuticle. Only older larvae devour entire terminal leaves. Initially, first instar larvae on male holly trees feed on the unopened flower buds.

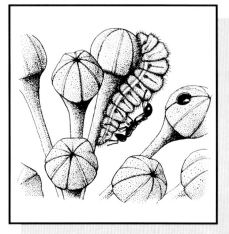

Figure 3: The larva on ivy with a Myrmica sp. ant in attendance

Second brood larvae on the ivy feed in a similar manner to those of the first generation on female holly trees. They burrow inside a developing flower bud, eating the contents. It is sometimes wrongly quoted that the larva feeds on the ivy berries. A small, neat, round entry hole in a flower bud indicates their former presence, but beware as some moth larvae also feed on the flower buds of ivy. Sprays of ivy flowers often have multiples of hatched eggs and therefore become extensively damaged, especially in good Holly Blue years. Such well utilised florets shrivel and dry. Also large collections of larval frass gather, which could also indicate the presence of a nearby larva.

The larva of the Holly Blue has three colour forms. The most common is an entirely green variety, with the exception of its black, shiny head,   present in all three forms and usually visible only when the larva is on the move. A second form retains the green ground colour but has dorsal and lateral discontinuous maroon stripes.   The most attractive of the three forms has strong rose pink continuous dorsal and lateral stripes, not too dissimilar to the colourful form of the Small Copper (*Lycaena phlaes*) larva. A typical well populated bank of ivy can comprise 90% of the common green form (Merton Park, September 1985).

The larva changes its skin three times and has been found in situ for its final moult on the underside of an ivy leaf, moving off its food source, to complete its transformation. As pupation nears, the larva loses its green coloration and turns a muddy maroon colour before wandering off in search of a pupation site.

THE PUPA

The pupa has rarely been found in the wild. One recent find (January 1998) was by Hampshire branch members during a conservation work party on the Butterfly Conservation Reserve at Magdalen Hill Down. This was during a scrub removing operation, some of which had ivy attached. Two pupae were discovered in the depths of the ivy growth and is probably the first recorded instance of wild pupae being located. In the South London parkland habitat, from the location of recently emerged, limp adults, pupation in the depths of ivy growths (under leaf, woody growth, leaf debris, wall crevices, brickwork) has always been suspected. The first brood on holly is reported to pupate amongst leaf litter and debris at the base of the tree or bush. The changed colour of the larva, prior to pupation, gives it excellent camouflage whilst dropping or crawling to the ground. In captivity the larva pupates on either surface of mature leaves or twigs. The pupal stage lasts approximately 200-220 days (overwintering brood) from September to May. Pupae resulting from the spring generation last a considerably shorter period from June to July.

The pupa has a brown ground colour, speckled with black and although somewhat shiny in appearance, is covered with short bristly hairs.

CONSERVATION

Thankfully, this widespread (within its range) and sometimes common butterfly, is not a threatened species. It is a common species because it has a wide selection of larval foodplants throughout its range and is not too specific to any type of habitat. This is unlike some of our threatened and declining species, several of which are monophagous, for example the Pearl bordered Fritillary (*Boloria euphrosyne*) on *Viola* sp. and the Adonis Blue (*Lysandra bellargus*) on *Hippocrepis comosa*. Both species also have strictly defined and limiting ecological requirements. A regular coppice system is required by the Pearl-bordered Fritillary and well grazed calcareous grassland by the Adonis Blue.

To assess and plan a conservation strategy for the Holly Blue would be an almost impossible task, due to its differing urban and rural habitat requirements. Future problems that it is likely to encounter are anticipated to be localised and confined to extensive losses of its larval foodplant. This would be particularly significant where there are 'core' populations. There have been some setbacks over the years in the South London parkland and these can be related to its prospects in the wider countryside.

By far the biggest difficulty the Holly Blue has experienced during its many generations in the South London parkland has been the progressive loss of its most important second brood foodplant - the ivy. From the early days of this comprehensive study one has been aware of a pending problem and early measures were taken, by the prompting of our former Chairman, Dr Harold Hughes to correspond with the relevant local council, making them aware of the importance of the site and the problems envisaged plus recommending remedial measures.

As usual with such approaches, there was a sudden flurry of interest, concern and practical measures taken (planting of ivy). Years pass and interest wanes, new councils are elected and there are staff changes and financial restraints. All this means that habitat improvements and former correspondence is either completely forgotten or deposited in the archives, destined for very low priority if any at all.

Approximately 70% of the ivy growing on the walls and archway of the South London park has now either died or been removed since the early 1970's. There have been some efforts to reinstate some of the ivy, but it is often replaced by Virginia Creeper (*Parthenocissus*) which has no intrinsic value to the Lepidoptera, unless one is captive-rearing the beautiful Indian Golden Emperor silkmoth (*Loepa katinka*)!

In the local churchyard, which is only some hundred metres from the 'core' population and also regularly inhabited by the Holly Blue, there has also been a quantity of ivy removed from old tombs, gravestones and along the boundary wall. The church has just celebrated its eight hundred and eighty third anniversary. The Holly Blue has perhaps always inhabited the churchyard and should be encouraged to continue its presence, although without its valuable second brood larval foodplant, it will become a passing visitor rather than the more desirable established churchyard resident.

It is unfortunate that the ivy has been persecuted for damaging brickwork and killing trees, especially in an urban environment. It is now well-known that ivy does not 'strangle' trees and only becomes a problem, particularly during stormy weather, because of the sheer volume of its weight. In the South London park a curved section of the ivy clad south facing wall toppled and fell during the winter of 1981 - 1982 due to the combined weight of snowfall and ivy growth. Ivy is also generally trimmed along some walls, where it has made significant summer growth and this could cause the loss of a small percentage of pupae when executed during the winter months, but is a much better operation than

14

complete removal. Such an operation will also reduce flower production the following summer.

The holly is not so much persecuted, but in some years has been heavily trimmed and cut (including the tall crowns) by council employees during maintenance procedures. This does little or no damage to the Holly Blue population, as it is undertaken during the winter months whilst the ivy is being utilised and overwintering pupae in situ. The leaf litter and debris at the base of holly trees and bushes, within a radius of several metres from the trunk, should remain undisturbed from May to July, to prevent damage to pupae awaiting emergence in the latter month. In the spring, holly flower production and trees of mixed sexes are desirable targets to encourage the Holly Blue, whilst autumn flower production on the ivy is equally important, both in the urban and rural environment. A southerly aspect is also preferred.   In the wider countryside females deposit eggs on a more extensive range of plants and shrubs (especially the first brood) and their usually smaller and scattered populations rely less heavily on holly, although ivy, as indicated throughout this booklet, is of major importance.

CONCLUSIONS

An **urban** study of an individual butterfly species is a rarely undertaken task. There has been some excellent recent work, *Butterfly Ecology in an Urban Cemetery* (Freed 1997, PhD. unpublished) and *Butterflies of Greater Manchester* (Hardy 1998), both in which the Holly Blue figures prominently and both contribute greatly to butterfly conservation in an urban environment.

The history of Merton Park is a valuable contribution to our understanding of how 'core' populations of Holly Blue can develop. Ivy should be promoted as a valuable larval foodplant for the Holly Blue. Churchyards are a very advantageous habitat, in which the larval foodplant is very frequent.   Many other species of butterfly also occur in churchyards, thus increasing their conservation value. A better understanding of Holly Blue pupation behaviour in the wild, would assist our ability to preserve populations of this species. Success rates in the wild of alternative larval food sources would also be useful.

Many butterfly species rely on the activities of man for their continued existence, not only in the British Isles, but throughout Europe. Man has grazed the downs with his domestic stock, providing the warm microclimates at ground level necessary for species on the edge of their range to flourish. Trees are felled and scrub coppiced thus providing the early successional plants for the species of more open woodland. A cessation of such anthropogenic activities, often due to financial restraints, usually causes problems for those species with limited ability to utilise a wide range of habitats. The Holly Blue is perhaps the best example of a species able to thrive in a man-made and heavily populated environment without the constraints of a change in use of land determining a restricted range. The vagaries of the British climate can be clearly seen as the most significant factor in determining population levels in any given year. The important host-specific parasite *Listrodromus nycthemerus* is also controlled by climatic influences and the resulting drop in the number and ease of locating available Holly Blue larvae in which to inject its eggs.

A definitive work on the ecology and conservation of the Holly Blue still eludes us. There is still much to be learned through careful observation, monitoring, and recording.

APPENDIX I

List of possible larval foodplants Holly Blue observed depositing eggs on during early 1990's years of abundance. (Not all listed below have evidence of larvae having actually fed and successfully completed their cycle on the foodplant).

Lucerne	*Symphoricarpus*	Sanfoin
Raspberry	*Buddleia*	Spindle
Blackberry	Hawthorn	Gorse
Purple Loosestrife	*Euonymus japonicus*	*Escallonia*
Pyracantha	Travellers Joy	Black Medick
Cornus sp.	Shrubby Senecio	Birds-foot Trefoil
Cotoneaster	Hemp Agrimony	Broom
Privet	Rowan	*Ceanothus*
Hebe sp.	Rosebay Willowherb	Alder Buckthorn
		Box

APPENDIX 2   Holly Blue Transect Data

TABLE 2 - Durlston Country Park (Eastern Sector SZ 0377)

YEAR	SPRING	SUMMER	YEAR TOTAL
1988	6	14	20
1989	9	48	57
1990	32	52	84
1991	26	54	80
1992	19	17	36
1993	2	3	5
1994	-	2	2
1995	4	13	17
1996	11	54	65
1997	44	86	130

TABLE 3 - Noar Hill (SU 741320)

YEAR	SPRING	SUMMER	YEAR TOTAL
1988	-	-	
1989	37	38	75
1990	93	177	270
1991	27	180	207
1992	1	4	5
1993	0	1	1
1994	0	1	1
1995	5	6	11
1996	31	43	74
1997	10	67	77

Holly Blue Transect Data

TABLE 4 - Ashford Hangers (SU 743269)

YEAR	SPRING	SUMMER	YEAR TOTAL
1988	1	1	2
1989	15	16	31
1990	12	19	31
1991	2	11	13
1992	0	2	2
1993	0	0	0
1994	0	0	0
1995	0	1	1
1996	4	9	13
1997	4	11	15

TABLE 5 - Wendleholme (SU 494073)

YEAR	SPRING	SUMMER	YEAR TOTAL
1988	3	5	8
1989	59	65	124
1990	59	83	142
1991	6	22	28
1992	3	5	8
1993	-	-	-
1994	0	0	0
1995	0	0	0
1996	21	23	44
1997	16	29	45

REFERENCES:

Chatfield, J 1987 - FW Frohawk - his life and work *Crowood*

Frohawk, FW 1934- The Complete Book of British Butterflies *Ward Lock & Co. London*

Frohawk, FW 1938- Varieties of British Butterflies *Ward Lock & Co. London*

Hall, ML & Buckley KL 1996 - A Management Plan for the Butterflies of Kew Gardens.

Heath, Pollard, Thomas 1984- Atlas of Butterflies in Britain and Ireland *Viking, Middlesex*

Harley Books 1989 - The Moths and Butterflies of Great Britain and Ireland : Volume 7

Hicken, N 1992 - The Butterflies of Ireland *Roberts Rinehart*

Jowett, EM 1951 - A History of Merton and Morden *Festival of Britain Local Committee*

Kawazoe, E 1983 - Blue Butterflies of the Lycaenopsis Group *British Museum Natural History*

Pollard, E & Yates, TJ 1993 - Monitoring Butterflies for Ecology and Conservation *Chapman & Hall*

Thomas, JA & Lewington, R 1991 - The Butterflies of Britain and Ireland *Dorling Kindersley, London*

Tolman, T 1997 - Collins Field Guide to the Butterflies of Britain and Europe *Collins*

THE AUTHOR

Ken Willmott FRES was born in 1949 and eleven years later became interested in butterflies whilst in the early years of senior 'Secondary Modern' school. The Holly Blue used to regularly pass through the school playground, which was in the heart of the Merton Park 'core' population.

Together with four other species texts, he wrote on the Holly Blue in volume seven of *'Moths and Butterflies of Great Britain and Ireland* (1989) by Harley Books of Colchester. He has also been commissioned to write on the Holly Blue for the forthcoming *'Butterflies of Hampshire'*.

18

Ken became a Fellow of the Royal Entomological Society in 1995 and has sat on the Conservation Committee of Butterfly Conservation for many years. He also gives talks regularly on a wide variety of butterfly subjects and has lectured on the ecology and conservation of British butterflies in many venues, including the University of Florida complex, Gainesville, USA as well as numerous locations in England.

Ken's colour transparencies have been used in numerous past and forthcoming publications and despite maintaining routine employment he continues to be eagerly involved in multifarious butterfly projects.

THE ARTIST

Dr Tim Freed is a freeelance illustrator and consultant in Lepidoptera site research. He achieved his BSc. at Aberdeen University and further studied natural history illustration at the Royal College of Art, where he obtained his PhD. in 1997 with illustrated research in Kensal Green Cemetery. This included studies on the Large, Small and Essex Skippers. Here he also studied the Holly Blue and the ecology of its larval foodplant - the Ivy. In 1992 he designed and created a butterfly garden in the Cemetery.

His artwork has been exhibited at the Natural History Museum in London and he has been one of the illustrators for *The Moths and Butterflies of Great Britain and Ireland* for Harley Books of Colchester since 1983.

ACKNOWLEDGEMENTS

I would like to acknowledge Tony Hoare, for typing the initial manuscript, many years ago, whilst he was the editor of Butterfly Conservation News. I have since expanded on the text which was to be included in an edition of the News. My thanks to Phil Grey of Dorset, Andy Barker of Hampshire, and Gail Jeffcoate of Surrey for the use of collated transect figures and Patrick Roper for his useful list of extracted notes from past journals. Thanks are also due to my wife Lorraine for her encouragement and patience, Carole Tucker for her invaluable help on the computer and Tim Freed for his superb and accurate drawings which enhance this booklet. Phil Budd for information on the wild pupa of the Holly Blue. Marney Hall for information gleaned from her Kew Gardens management plan. Pippa Howard for her professional skills on the Scanning Electron Microscope. Martyn Davies, Steve Jeffcoate, Howard Whiting and Simon Glover for making this project viable. I would also like to acknowledge Bill Gerard for his invaluable review of the text. Finally, thanks are due to all the other authors in this series of booklets for their inspiration to be involved with such a useful and collectable set of volumes.

BUTTERFLY CONSERVATION

BUTTERFLY CONSERVATION (the British Butterfly Conservation Society Ltd) is probably the largest insect conservation body in the world and is devoted to the conservation of our native butterflies and moths, and their habitats. Butterfly Conservation is dynamic, fast-growing and influential. Our activities span all spheres of modern conservation work.

Our ultimate goal is the restoration of a balanced countryside, with butterflies and other wildlife returned to the profusion they, and we, once enjoyed.

BUTTERFLY CONSERVATION

Is playing a leading role in conserving butterfly and moth populations, particularly through the preparation of Species Action Plans

◆ Campaigns to save threatened habitat

◆ Carries out research on threatened butterflies and moths

◆ Is a member of the influential bio-diversity challenge group

◆ Acquires and manages nature reserves

◆ Has branches covering the whole of the UK which promote conservation at a local level

◆ Surveys, records and monitors butterflies and moths throughout the UK

◆ Works in partnership with other conservation bodies

◆ Encourages an interest and awareness of butterflies, moths and their conservation

◆ Advises landowners on habitat management

◆ Promotes invertebrate conservation generally

◆ Publishes Butterfly Conservation News three times a year

◆ Lobbies National and Local Government to influence planning and policy decisions

BUTTERFLY CONSERVATION
P.O. Box 222 Dedham
Colchester Essex CO7 6EY

BUTTERFLY CONSERVATION 1999
THE BRITISH BUTTERFLY CONSERVATION SOCIETY , Shakespeare House, High Street, Dedham, Colchester, Essex, CO7 6DE registered Charity No 254937